ISAAC ASIMOV'S
LIBRARY OF THE UNIVERSE

THE BIRTH AND DEATH OF STARS

by Isaac Asimov

DELL YEARLING · NONFICTION

Published by
Dell Publishing
a division of
Bantam Doubleday Dell Publishing Group, Inc.
666 Fifth Avenue
New York, New York   10103

This edition was first published in the United States and Canada in 1989 by Gareth Stevens, Inc.

Technical advisers and consulting editors: Fran Millhouser, Julian Baum, and Francis Reddy

ISBN: 0-440-40446-0

Reprinted by arrangement with Gareth Stevens, Inc.

Printed in the United States of America
October 1991

10 9 8 7 6 5 4 3 2 1

CONTENTS

Nowadays, we have seen planets up close, all the way to distant Neptune. We have mapped Venus through its clouds. We have seen the rings around Neptune and the ice volcanoes on Triton, one of Neptune's moons. We have detected strange objects no one knew anything about until recently: quasars, pulsars, black holes. We have studied stars not only by the light they give out but by other kinds of radiation: infrared, ultraviolet, x-rays, radio waves. We have even detected tiny particles, called neutrinos, that were given off by an exploding star in another galaxy.

We have come to understand that the Universe changes. It may not seem so to us because the changes are so slow. However, stars come into being and are born; they change and grow older; and eventually, they come to an end, sometimes in violent ways. In this book, we will consider how stars are born and how they die.

Isaac Asimov

Out of the Darkness

About 15 billion to 20 billion years ago, out of a darkness as deep and pitch-black as any you can imagine, the Universe burst into being.

We call this beginning the Big Bang. As the Universe expanded and grew vast, two kinds of atoms were formed — hydrogen and helium. These were the simplest atoms. There were enormous clouds of these gases. These broke up into larger and smaller fragments. Out of them were formed the galaxies and the many billions of stars that make up those galaxies.

Not all clouds formed stars immediately. In fact, it wasn't until nearly five billion years ago that a particular cloud of gas came together to form our own Sun.

Opposite: the Big Bang — the tremendous cosmic explosion that created our Universe.

Below: the history of the Universe. The bright spot at left represents the Big Bang, while the ruler represents the growing size of the Universe and the watch represents passing time. As you look further in time to the right, the Universe grows larger, cools, and eventually forms the stars and galaxies we see today.

Children of the Nebula

A large cloud of gas has a gravitational pull and a turning motion. Its gravitational pull causes it to begin condensing, or coming together. As it does so, it turns more rapidly and grows warmer. It condenses faster and still faster, until it forms a protostar, which is an early version of a star.

Not all clouds are large enough to start condensing on their own. Some may start collapsing when a nearby star that has already formed explodes. The force of the blast pushes the cloud matter together, and then the cloud continues to come together on its own.

The galaxies — lumps in space?

Some scientists think that at the Big Bang, the Universe was an evenly spread mass. As the Universe expanded, it should have stayed even in all directions and become one big, even cloud of gas. That didn't happen. Instead, the gas broke up into galaxy-sized clouds that slowly collapsed into stars. This has produced a "lumpy" Universe, but scientists are not sure how that lumpiness came to be. We face a mystery right from the very beginning of the Universe!

Opposite: Out of this cool, dark gas cloud somewhere in the Galaxy, a star has begun to form out of a clump of gas.

Far opposite: The young star grows as its gravity gathers more gas from the surrounding cloud.

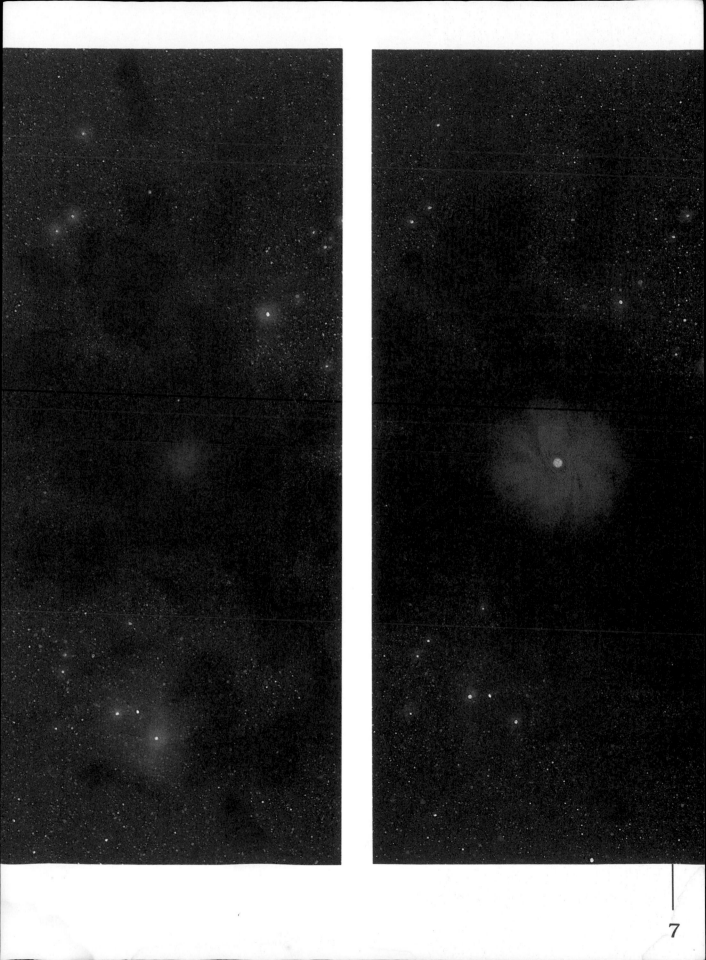

A Star Is Born

The gas cloud takes millions of years to come together, all the time getting denser, or more tightly packed, and hotter. At the center of the cloud, the matter is densest and hottest, and the atoms there slam together harder and harder as a result.

Eventually, when the temperature reaches tens of millions of degrees, hydrogen atoms smash together to form the more complicated helium atoms. This process, called nuclear fusion, releases a great deal more heat, and the center of the cloud glows white hot. It has become a star.

The planets in our Solar system were formed in a similar manner. The thinner matter in the outer regions of the Solar system came together and formed the planets circling our Sun.

Right: More gas falls onto the young star, compressing it and raising the temperature of its central core. Eventually, the temperature becomes high enough for energy-producing reactions to take place. A new star is born!

Opposite: The blue stars lighting up the gas in the Rosette Nebula may be only half a million years old — young stars!

**Multiple stars —
not as odd as you might think**

Often a cloud of gas will collapse —
not into a star, but into two or more
closely spaced stars. When such
multiple stars were first discovered,
they were thought to be rare. As
time went on, they turned out to be
more and more common. Now we
think that at least half of all stars
are multiple. What makes a cloud
condense into a single star or into
multiple stars, and how does that
influence the formation of planets?
As yet, we simply don't know.

Our Sun —
A Very Ordinary Star

Stars come in all sizes. Our Sun is about average in size and is quite ordinary. Once a star forms, the heat it produces by nuclear fusion expands it enough to balance the contraction of its gravitational pull. The star remains stable and doesn't change much for millions of years.

Our Sun has been shining for some 4.5 billion years and is only middle-aged! It will keep on shining as it is for about five billion years more. The hydrogen at its center turns slowly to helium, and the heat produced gives us light and warmth and makes it possible for life to exist on Earth.

Above: The single star on which we depend seems to be unusual. Most stars are born with partners.

Opposite: a stellar "nursery." The contracting gas of a star-forming cloud triggers the birth of thousands of stars, usually in clusters. It's rare when a star forms alone, far from the influence of other new stars.

Right: the Great Nebula in Orion. This great gas cloud is just the visible surface of a far larger star-forming cloud. Here the newborn stars light up the gas as they emerge into the Galaxy.

Giants in the Night

Most stars are both smaller and cooler than the Sun, shining with a dim red light. A few are larger, hotter, and more luminous than our Sun. The two brightest stars of the constellation known as the Southern Cross are thousands of times more luminous than our Sun. Beta Centauri is over 8,200 times as luminous, and the star Rigel (pronounced RI-jel), which is in the constellation Orion, is over 52,000 times as luminous as our Sun.

To remain so bright, very large stars must use their hydrogen fuel very quickly. Even though their large sizes give them a huge supply, such stars don't last long. They shine for only a few million years before they use up their hydrogen.

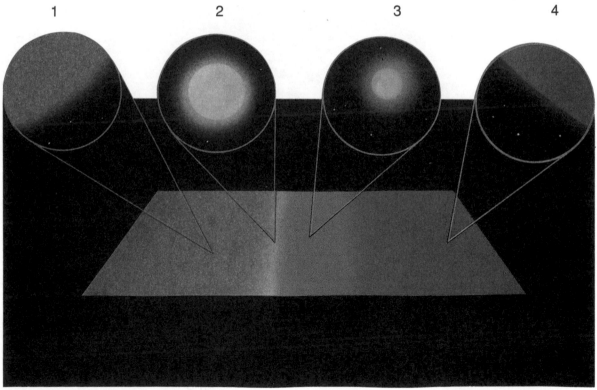

The color of the light a star produces depends on its mass, age, and temperature. Hot, young, and massive blue-white stars like Lambda Cephei (1) and Sirius (2) burn their fuel quickly. Our yellow Sun (3), a relatively small, cool star, burns its fuel slowly. A massive star near the end of its life, like Betelgeuse (4), glows a cooler red but gives off great heat because of its large surface.

With the help of an open camera lens and Earth's rotation, a night of winter skywatching shows that different types of stars have different colors caused by different stellar temperatures. The star trails (identified in the diagram) display a range of pale, barely detectable colors: Procyon (yellow-white), Pollux (orange), Castor (blue-white), Aldebaran (orange), red giant Betelgeuse (red-orange), and the brightest nighttime star, Sirius (white).

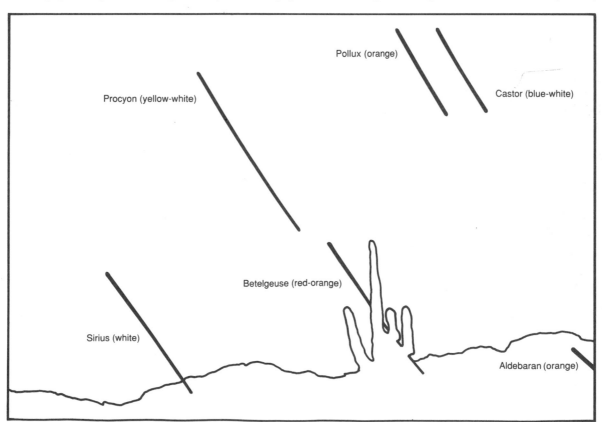

Red Giants

While a star is stable and shines steadily, it is on what we call the "main sequence," just as our Sun is now. Eventually, though, more and more helium collects in its center. As its hydrogen fuel runs out, the star contracts and squeezes its helium. The central temperature rises until the helium atoms begin to form still more complicated atoms — carbon, oxygen, and even heavier elements like iron. The extra heat makes the outer layers of the star expand; the star grows larger and larger. As the outer layers expand, they become cooler and glow red hot. Such stars are called red giants.

As far as we know, all stars become red giants as their hydrogen runs low, but large ones become supergiants. Betelgeuse (BEE-tul-jooz), which is in Orion, is a red giant that might be as much as 700 million miles (1.13 billion km) across. That's more than 800 times as wide as our Sun!

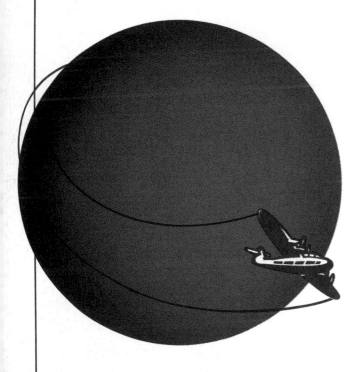

Flying at about 200 miles (320 km) an hour, an old-fashioned plane like the one shown here would need about 1 1/2 years to fly around our "small" Sun. But flying around a red giant like Betelgeuse would take over 800 times as long — about 1,200 years. All aboard!

Opposite: An artist imagines a planetary system near a red giant star. In a few billion years, our Sun will also bloat into a red giant.

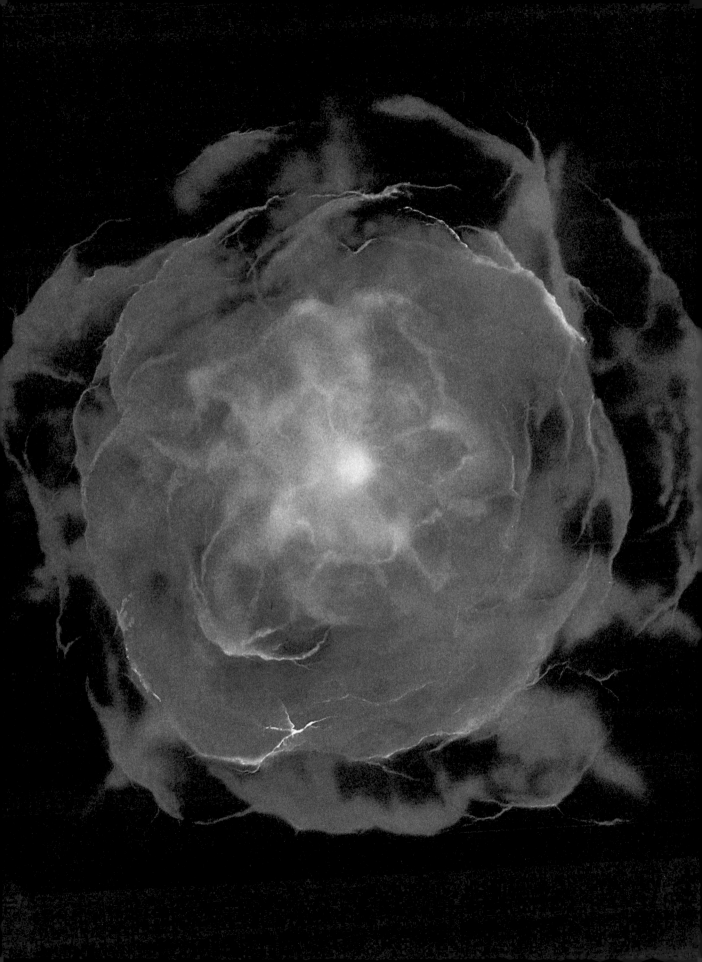

To Outshine the Galaxy

When enough atoms of iron and other heavy elements build up in a star's center, no more fusion can take place. The center cools off, and without the fierce heat to balance gravity, the star collapses. When this happens, the outer layers are blown off. The larger the star, the more violent the explosion.

Scientists believe that only really large stars — or small stars that attract matter from a nearby companion star — can blow up as what we call a supernova. During the explosion, the star's outer layers grow so hot that for a few weeks a single supernova will shine as brightly as an entire galaxy of stars.

Supernovas have happened in our own Galaxy, and witnesses say that some appeared in the sky as bright as the planet Venus! The last time this happened was in 1604, before the telescope was invented.

Opposite: an artist's concept of a supernova's expanding shell of gas. These stellar explosions — the death of one star — may trigger the birth of others by forcing nearby gas clouds to collapse.

The neatest trick of the week? — predicting a supernova

Supernovas always catch us by surprise. They're usually well under way before they're seen. If we could predict when a star might explode, scientists could prepare instruments to study the first moments of the explosion and even the time before it. But we can't tell when stars will explode. We know a star has to be in the red giant stage, but whether a particular red giant will explode tomorrow or 10,000 years from now, we just can't say — at least not yet. ●

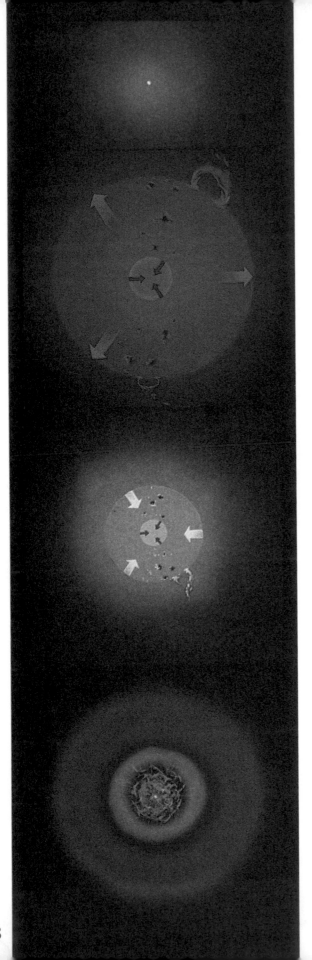

The Case of Supernova 1987A

Since 1604, the supernovas we have studied have occurred in distant galaxies and could only be seen dimly through telescopes. In February 1987, a supernova exploded in the Large Magellanic Cloud, a small galaxy only about 160,000 light-years from us. This was the closest supernova in nearly four centuries. It was about a billion times as bright as our Sun!

Supernova 1987A was the first since 1604 that could be seen by eye, but only in the Southern Hemisphere. For the first time, scientists could use modern instruments to study a supernova that was fairly close. They studied just how radiation and neutrinos were given off and how a cloud of gas formed and expanded.

As a massive star ages, it grows larger, cooler, and redder. When its internal fuel is finally exhausted, the whole star collapses — in less than a second! This triggers the tremendous explosion we call a supernova.

Top: The latest known supernova appeared to astronomers in 1987. The star exploded in a small galaxy that orbits our own. Here's what the area looked like before the explosion.

Bottom: The bright new star in this photo isn't new at all. It is the final burst of light from a large and unstable star. This photo was taken just after the explosion.

A look back in time

Although we <u>saw</u> a star explode in February 1987, that is not when it exploded. The supernova is 160,000 light-years from us. That means light from it takes 160,000 years to reach us. The explosion took place some 160,000 years ago, but the light from it just reached us in 1987. Some very distant objects are about 17 billion light-years away. We see them as they looked 17 billion years ago. That's how long it took their light to reach us.

Ghosts of the Giants

As a star like our Sun collapses, gravity breaks atoms into smaller pieces and forces them close together. The whole mass of a star can be squeezed into an object about the size of a planet like Earth! This shrunken star shines white hot and is called a white dwarf.

A still larger star collapses more violently, squeezing the pieces of atoms. After a supernova explosion dies down, what was once a red giant becomes a tiny, tightly packed ball perhaps only <u>eight</u> miles (13 km) across. This is a neutron star.

Squeezed even <u>more</u> tightly together, the center of a collapsing star could become a black hole.

1   2   3

Right and below: One of the stars in this binary system exploded long ago, leaving behind a dense neutron star. Its companion, now an aged red giant, loses part of its atmosphere to the intense gravity of the neutron star. This fresh gas will let the neutron star shine again.

Opposite and below: Imagine that you could scoop up and weigh a cup of matter from various kinds of stars. You'd be surprised at the results!

Weighing a scant 0.00002 pound (0.000009 kg), a cup of average red giant matter (1) would barely lower the scale. A cup of matter from a star like our Sun (2) would tip the scale at 0.73 pound (0.333 kg). Your scale would crumble under a 5.1-million-pound (2.3-million-kg) cup of white dwarf matter (3). And a cup of neutron star matter (4) would register an amazing 730 <u>trillion</u> pounds (331 trillion kg) — if you could find a scale to weigh it on!

Neutron stars — fast and deadly

Neutron stars spin very rapidly, often many times a second. Recently astronomers have found some neutron stars that rotate over 600 times each second. They have also found neutron stars that circle ordinary stars. With their powerful gravitational pull, the neutron stars slowly suck the material of those ordinary stars into themselves and destroy them. Only in the last few decades have we discovered what an active Universe we inhabit.

4

The Death of Earth

Someday our Sun will die, too, or at least it will cease being the kind of star it is now. About five billion years from now, it will start consuming helium in its core and it will start expanding and turning red. The Sun's surface will be cooler than it is now, but as it expands there will be so much surface that the amount of heat altogether will become far greater.

Life on Earth will become impossible. Eventually, in fact, the Sun will expand so much that it will swallow up Mercury, Venus, and then Earth. In time, it will engulf even Mars. By that time Earth will have turned into vapor. Our planet will no longer exist.

Life on Earth developed as our Sun settled into its current stage (opposite, left). In a few billion years, when the Sun expands into a red giant, things will change quite a bit! First, the ocean water will turn into vapor, creating thick clouds (opposite, right).

In time, Earth's oceans will boil away. Our planet's crust will soften and begin to melt (left). Before Earth becomes lost in the growing outer layers of the Sun, our neighboring Moon will dissolve (above).

From Red Giant to White Dwarf

Although Earth will be gone, the Sun will still be there as a red giant. Less than a billion years after it becomes a red giant, the Sun will collapse. It will be too small to collapse violently enough to become a supernova. It will become a white dwarf, scattering its outermost layers in all directions and forming a shell of gas called a planetary nebula. We can see some of these in the sky — a central white dwarf shining within a shell of expanding gas.

The Sun will continue to shine as a white dwarf for billions of years, until, like a burned-up cinder, it gradually cools into a black dwarf.

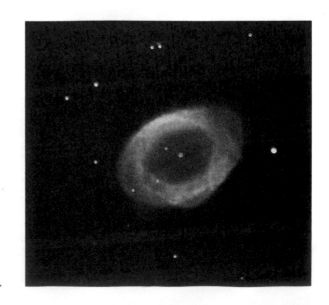

Ring nebulas are gaseous shells lit up by white dwarfs, the final remains of Sun-like stars. Because of their disklike appearance in telescopes, these clouds are called planetary nebulas. Here is a gallery of planetary nebulas.

Right: Ring Nebula in the constellation Lyra.

A red giant as seen from a planet at a "safe" distance. Groups of sunspots dot its ruddy surface, and giant flares loop gas filaments far into space.

The Bug Nebula in Scorpius.

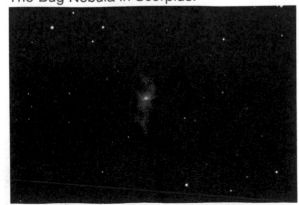

The Eskimo Nebula in Gemini.

A New Beginning

Out of the deaths of stars may come the births of other stars.

When a supernova explodes, most of its matter is scattered through space. This matter contains atoms more complicated than those of hydrogen and helium — like carbon, oxygen, silicon, and iron. The clouds of gas that still exist between the stars therefore come to contain these complicated atoms. New stars form out of these clouds, and they too contain the complicated atoms.

Because these stars are "born" out of matter from earlier stars, they are called second-generation stars. Our Sun is an example of a second-generation star. Earlier stars could only have planets made up of hydrogen and helium, somewhat like Jupiter. Only second-generation stars can have planets made of rock and metal, like Earth.

And since the complicated atoms that make up our bodies also had their origins in ancient stellar matter, we, too, are "born" out of the stars!

Children of the stars

After the Big Bang, only hydrogen and helium atoms formed. All atoms more complicated than these formed later in the centers of stars. These atoms were then spread throughout space by giant supernova explosions. Our planet Earth and most of our own bodies are made up almost entirely of these complicated atoms. Almost every atom within our planet and ourselves was formed at the center of some star that once exploded violently. We are children of the stars.

The next time you take a stroll beneath the Milky Way, imagine the stars as your cosmic birthplace.

Fact File: Different Stars, Different Paths

All stars are born in pretty much the same way. But their sizes and masses affect how long they live — and how they die. The smallest stars are only about eight miles (13 km) across — while the largest ones are millions of times bigger than our Sun. Using the chart on these two pages, let's follow the lives of three "typical" stars — a giant star, a star about the size and mass of our Sun, and a star about a third of the mass of our Sun.

Time from Nebula Stage	Giant Star (15 solar masses)	"Average" Star (one solar mass)	Small Star (one-third solar mass)
10,000 years	Matter contracts into large ball; resulting giant star has a lot of matter, but burns very quickly.		
1 million years		Matter contracts into hot core; protostar forms — about 20 times Sun's brightness and width.	
10 million-11.6 million years	Helium created by nuclear fusion begins to build up in core. As the star feeds on its helium, it expands to as much as 100 times the diameter of our Sun.		Temperature inside the core of the nebula that will give birth to a red dwarf is just enough to start burning. Low temperature means future red dwarf star will burn its hydrogen slowly and "live" longer than larger stars.
11.9 million years	The star expands into a red supergiant. As the massive star continues to burn, its outer layers explode — a supernova! At the center of the explosion, a tiny — perhaps only eight miles (13 km) wide — but incredibly massive neutron star is left.		

After a "short" life of a few million years, the 15-solar-mass giant star has burned itself out. But the other two stars have barely begun their lives. Their story continues below.

Time from Nebula Stage	"Average" Star (one solar mass)	Small Star (one-third solar mass)
70 million years	Hydrogen atoms in protostar begin the fusion process. Protostar becomes a star.	
1 billion years		Red dwarf now fully formed. Average surface temperature might reach only about 4900°F (2700°C) — cool for a star.
4.5 billion years	The star is now just like our Sun.	
7 billion-9 billion years	Star begins to run out of hydrogen. As it burns remaining hydrogen, its temperature and size increase.	
10 billion-11 billion years	Star begins to consume its helium and balloons into a red giant. By now, any planets once suitable for life — like Earth — have long been charred, lifeless wastelands. At its largest, this star is 400 times the diameter of the Sun — big enough to swallow Mars in its orbit!	The giant star has been "dead" for about 10 billion years. The star like our Sun is starting to explode into a red giant. But the red dwarf, slow, steady, and stable, is still just starting its life!
11 billion-11.7 billion years	As star uses up its fuel, it shrinks into a white dwarf only about as big as Earth.	
50 billion-100 billion years?	As far as we know, the Universe is only 17-20 billion years old. But we think that the white dwarf may burn out completely in 50 billion years — and become a black dwarf.	Red dwarf stars may burn on quietly for some 100 billion years. They might become tiny white dwarfs or "brown dwarfs." No one really knows for sure!

More Books About the Stars

Here are more books about the birth, life, and death of stars. If you are interested in them, check your library or bookstore.

Bright Stars, Red Giants and White Dwarfs. Berger (Putnam)
How Was the Universe Born? Asimov (Gareth Stevens)
Our Solar System. Asimov (Gareth Stevens)
Quasars, Pulsars, and Black Holes. Asimov (Gareth Stevens)
Stars and Galaxies. Apfel (Franklin Watts)
The Stars: From Birth to Black Hole. Darling (Dillon)
The Sun. Asimov (Gareth Stevens)
Sun and Stars. Barrett (Franklin Watts)

Places to Visit

You can explore the remote reaches of the Galaxy and discover the wonders of stellar birth, life, and death without leaving Earth. Here are some museums and centers where you can find a variety of space exhibits.

Adler Planetarium
Chicago, Illinois

Hayden Planetarium
Boston, Massachusetts

Kansas Cosmosphere and Space Center
Hutchinson, Kansas

NASA Lyndon B. Johnson Space Center
Houston, Texas

H. R. MacMillan Planetarium
Vancouver, British Columbia

Touch the Universe — Manitoba Planetarium
Winnipeg, Manitoba

For More Information About the Stars

Here are some places you can write to for more information about the stars. Be sure to tell them exactly what you want to know about or see. Remember to include your age, full name, and address.

For information about stars:
Star Date
McDonald Observatory
Austin, Texas 78712

For photographs of stars:
Caltech Bookstore
California Institute of Technology
Mail Code 1-51
Pasadena, California 91125

NASA Jet Propulsion Laboratory
Public Affairs 180-201
4800 Oak Grove Drive
Pasadena, California 91109

For monthly sky maps:
National Museum of Science and Technology
Astronomy Division
P. O. Box 9724
Ottawa Terminal K1G 5A3, Canada

Glossary

the Big Bang: a gigantic explosion that some scientists believe created our Universe.

billion: in North America — and in this book — the number represented by 1 followed by nine zeroes — 1,000,000,000. In some places, such as the United Kingdom (Britain), this number is called "a thousand million." In these places, one billion would then be represented by 1 followed by *12* zeroes — 1,000,000,000,000: a million million, a number known as a trillion in North America.

black dwarf star: a "dead" star. When a star like our Sun uses up its store of hydrogen energy and collapses, it becomes a shining white dwarf star. When a white dwarf stops giving off light, it becomes a black dwarf.

black hole: a massive object — usually a collapsed star — so tightly packed that not even light can escape the force of its gravity.

galaxy: any of the many large groupings of stars, gas, and dust that exist in the Universe.

helium: a light, colorless gas that makes up part of every star.

hydrogen: a colorless, odorless gas that is the simplest and lightest of the elements. Most stars are three-quarters hydrogen.

infrared radiation: "beneath the red" radiation. Infrared wavelengths are longer than red light wavelengths. Infrared is a form of invisible light, but you can feel infrared as heat.

luminous: producing light.

main sequence: a class of stars that shows a stable relationship between brightness, size, and temperature; the "middle age" of a star's life — the age into which our Sun falls.

nebula: a cloud of dust and gas in space. Some large nebulas, or nebulae, are the birthplaces of stars. Other nebulae are the debris of dying stars.

neutrino: a subatomic particle produced when hydrogen fuses to helium in the center of a star.

neutron star: a star with all the mass of an ordinary large star but with its mass squeezed into a much smaller ball.

nuclear fusion: the smashing together of highly heated hydrogen atoms. This fusion of atoms creates helium and produces tremendous amounts of energy.

planetary nebula: a shell of gas expelled by a red giant star that has used up much of its hydrogen fuel, leaving the star's core as a brightly shining white dwarf.

proto-: the earliest or first form of something. In this book, we talk about a young star as a "protostar."

pulsar: a neutron star with all the mass of an ordinary large star but with that mass squeezed into a small ball. It sends out rapid pulses of light or electrical waves.

radiation: the spreading of heat, light, or other forms of energy by rays or waves.

radio waves: electromagnetic waves that can be detected by radio receiving equipment.

red dwarf star: a cool, faint star, smaller than our Sun. Red dwarfs are probably the most numerous stars in our Galaxy, but they are so faint that they are extremely difficult to see.

red giant: a huge star that develops when its hydrogen runs low and the extra heat makes it expand. Its outer layers then change to a cooler red.

supernova: a red giant that has collapsed, heating its cool outer layers and causing a huge explosion.

white dwarf: the small, white-hot body that remains when a star uses up its store of nuclear energy and collapses.

Index

The publishers wish to thank the following for permission to reproduce copyright material: front cover, p. 9, © Rick Sternbach; p. 4, © Julian Baum, 1988; p. 5, © Mark Paternostro, 1988; pp. 7 (both), 8, 21, Paul Dimare, 1988; p. 10 (lower), Matthew Groshek/© Gareth Stevens, Inc.; pp. 10-11 (upper), © Matthew Groshek, 1980; pp. 11 (lower), 19 (both), 24, 25 (bottom left), National Optical Astronomy Observatories; p. 12, © Brian Sullivan, 1989; p. 13 (upper), © Allan Morton ; p. 13 (lower), Sharon Burris/© Gareth Stevens, Inc.; p. 14, Kate Kriege/© Gareth Stevens, Inc.; p. 15, © Paul Dimare; p. 16, © Sally Bensusen/Davis Planetarium, 1987; p. 18, © Sally Bensusen, 1988; pp. 20-21, © Lynette Cook, 1988; pp. 22-23 (all), 25 (upper), © John Foster; p. 25 (bottom right), © 1984/University of Hawaii Institute of Astronomy, by Jack Marling and Wayne Annala, 24" f/15 telescope; p. 26, © Greg Mort, 1987.